AF702751

INTRODUCTION

This is the forty-third year of publication for the Earthworm Buyer's Guide and Directory. This new edition has been edited to give you the most accurate list yet compiled of active, leading worm growers. Changes in listings are occurring constantly so occasionally you may find an individual has moved or is no longer a worm grower.

All the growers are listed by states in the General Listings. There are additional listings according to different kinds of worms plus listings of miscellaneous items. These are by name, city and state only. Full addresses and detailed information can be obtained from the General Listings and Display Ads.

Charlie Morgan, whom we consider the leading authority on earthworm farming, offers a word of advice to those considering the worm business as a money-making project. Start with a modest investment and make sure you know from hands-on experience that you can raise worms. Let the business begin to pay for itself by producing, packing and selling on a small scale the first year. To be successful in raising and selling worms on a commercial scale requires hard work. This advice would apply to any new venture, but we want to emphasize it to newcomers in worm farming.

To those already established in worm farming who would like to be listed in the next edition of the Earthworm Buyer's Guide, we will be glad to send details.

SHIELDS PUBLICATIONS
P.O. Box 669
Eagle River, WI 54521

GENERAL LISTINGS

ALABAMA

Robert E. Tate
Tate's Worm & Fern Farm
Rt. 1, Box 503
CHATOM, AL 36518
Phone: 205-847-2976
Red Worms, Brown Nose Nightcrawlers, Grey Wigglers
Mature, Bait, Pit-run
Mail order
Wholesale & Retail
(See display ad pg. 44)

* * *

Ezell Hall
Hall's Organic Gardens
1805 Colvert St.
TUSKEGEE INSTITUTE, AL 36088
Phone: 205-727-6365
Red Worms; Mature, Pit-run
Organically grown plants, potting soil, compost
Wholesale & Retail

* * *

ARKANSAS

Glenn Foster (Linda)
Worms by Weber
P.O. Box 1126
YELLVILLE, AR 72687
Phone: 501-449-5348
Red Worms, African Nightcrawlers, Tiger Wigglers
Mature, Bait, Pit-run
Castings, Worm School, Organic Garden & Lawn Supplies
Mail order & Local;
Wholesale & Retail
(See display ad pg. 58)

* * *

Jim Knight
Rt. 2, Box 183
EUDORA, AR 71640
Phone: 501-355-2154
Brown Nose Worms;
Mature, Bait;
VCR Tapes, Shipping Bags, Tumblers;
Mail Order
Wholesale & Retail
(See display ad pg. 44)

Ed & Pat Griffin
Griffin Farms
8401 Crossroads
BENTON, AR 72015
Phone: 501-776-1750
Red Worms, African Nightcrawlers, Arkansas Tiger Stripes;
Mature, Bait, Pit-run;
Worms Castings
Mail Order & Local
Wholesale & Retail
(See Display ad pg. 42)

* * *

John W. Powell
Powell's Bait Farm
Star Rt., Box 18
WISEMAN, AR 72587
Phone: 501-322-7344
Red Worms; Mature, Bait;
Mail Order & Local;
Wholesale & Retail
(See display ad pg. 48)

* * *

Hilda Stewart
Family Worm Farm
1553 Greene 611 Rd.
PARAGOULD, AR 72450
Phone: 501-586-0451
Red Worms; Mature, Pit-run;
Mail order;
Wholesale
(See display ad pg. 52)

* * *

J. & Scott Downs
J&S Worm Farms
411 S. 4th Ave.
PARAGOULD, AR 72450
Phone: 501-236-7619
Red Worms; Mature, Pit-run;
Worm Bags;
Mail order & Local;
Wholesale & Retail
(See display ad pg. 50)

Dan Mechura, Sr.
National Rabbit Enterprises
P.O. Box 913
ROGERS, AR 72757
Phone: 501-636-8376
Catfish Bait Worms (Swamp Nightcrawlers), Red Worms; Mature;
Worm Books, Rabbit Books, Supplies
Wholesle & Retail

* * *

John Loseman Jr.
John's Worm Ranch
315 Mt. Zion Lane
BATESVILLE, AR 72501
Phone: 501-793-5808
Red Worms; Mature, Bait; Pit-run;
Earthworm Books
Mail Order & Local
Wholesale & Retail

* * *

Joan Hartfiel
Rt.2, Box 155
PARIS,AR 72855
Phone: 501-963-3282
Red Worms; Mature, Bait, Pit-run;
pH Testers, Earthworm Books,
Worm Harvesters
Wholesale

* * *

Phillip O. Sapp
Red-Eye Bait Farm
P.O. Box 4464
FT. SMITH, AR 72914-4464
Phone: 501-783-2952
Red Worms, Native Nightcrawlers;
Mature, Bait, Pit-run;
Mail Order & Local
Wholesale & Retail

* * *

JoAnn Filerbo
Filerbo Worm Farm
193 Greene 619 Rd.
PARAGOULD, AR 72450-8500
Phone: 501-586-0568
Red Worms; Mature, Pit-run;
Mail Order, Wholesale

* * *

Steve Hames
Wholesale Worms
203 Lacy St.
SILOAM SPRINGS, AR 72761
Phone: 501-524-8442
Red Worms;
Mature, Bait, Pit-run;
Book and information;
Mail Order & Local
Wholesale & Retail

* * *

CALIFORNIA

Hy Hunter
Hy Hunter Industries
12001 Highwater Rd.
GRANADA HILLS, CA 91344
Phone: 818-363-0202
Red Worms, Brown Nose,
African Nightcrawlers, Native
Nightcrawlers, Calif. Super
Worms from China;
All worm-growing supplies;
Mail Order & Local;
Wholesale & Retail
(See display ads pgs. 28,29,38)

* * *

Alfred D. Cardoza
Rainbow Worm Farm
24700 County Road No. 95
DAVIS, CA 95616
Phone: 916-758-9906 FAX no. 916-756-0414
Red Worms; Mature, Bait, Pit-run;
Harvesters, Bagging machine & sealer,
Mailing cartons, Potting soil &
castings, bulk & sacked;
Mail order & Local, export
Wholesale & Retail
(See display ad pg. 34)

* * *

Norbert Schwecke
Norb's Bait Farm
1370 2nd Street
NORCO, CA 91760
Phone: 714-737-3459
or 714-371-3310
Video Instructions
of Worm Farming
Mail Order & Local
(See display ad pg. 31)

CALIFORNIA CONT.

Fred Rhyme
Rainbow Mealworms Inc.
126 E. Spruce St.
COMPTON, CA 90220
Phone: 213-635-1494
Meal Worms, Crickets, Wax Worms;
Mail order & Local
Wholesale & Retail
(See display ad pg. 30)

* * *

Calvin R. Sierer
Wormco
P.O. Box 338
LINCOLN, CA 95648
Phone: 916-645-9419
Red Worms, Mealworms; Mature,
Bait, Pit-run;
Castings, manure, starter soil;
Mail order & Local;
Wholesale & Retail
(See display ad pg. 59)

* * *

James A. Cain
Cain's Farm
5678 Weber Rd.
DIXON, CA 95620
Phone: 707-448-6836
Red Worms; Mature, Bait,
Pit-run; Castings;
Mail Order & Local
Wholesale & Retail
(See display ad pg. 59)

* * *

Carl Denton
Denton's Quality Worm Farm
6441 Byrnes Rd.
VACAVILLE, CA 95687
Phone: 707-448-0453
Red Worms, Gray Crawlers (Jumpers);
Mature, Bait, Pit-run;
Castings;
Mail Order & Local
Wholesale & Retail
(See display ad pg. 50)

Richard Morhar
The Worm Concern
1450 Tierra Rejada Rd.
SIMI VALLEY, CA 93065
Phone: 805-520-1150
Red Worms, Canadian Nightcrawlers;
Mature, Bait, Pit-run; Castings,
Worm Boxes, Organically grown plants,
Organic recycling, Soil;
Mail order & Local
Wholesale & Retail

* * *

Tim & Peggy Matthews
Redwood Valley Worm Farm
6196 Redwood Valley Rd.
BLUE LAKE, CA 95525
Phone: 707-668-5997
Red Worms; Mature, Bait, Pit-run;
Earthworm Castings
Mail Order & Local
Wholesale & Retail

* * *

Mr. Marion Person
Aarion Wormery I
2195 Spencer Ave.
POMONA,CA 91767
Phone: 714-623-0714
Red Worms; Mature & Pit-run;
Mail Order & Local
Wholesale & Retail

* * *

Glenn Davis
Little Critter's Worm Farm
P.O. Box 216
NORTH PALM SPRINGS, CA 92258
Phone: 619-954-7444
Red Worms; Mature, Bait, Pit-run;
Mail Order & Local;
Wholesale & Retail

* * *

Philip Snow
Earthworks Enterprises
4691 Suisun Valley Rd.
SUISUN,CA 94585
Phone: 707-864-1701
Red Worms; Mature, Bait,
Pit-run; Castings;
Mail Order & Local;
Wholesale & Retail

Kurt E. Papenfuhs
7384 Condalia Ave.
YUCCA VALLEY, CA 92284
Phone: 619-365-7552
Jobber of Crickets,
Red Worms, Starter kits;
Local; Wholesale

* * *

Jonne Stauffers
The Potted Thorne Nursery
P.O. Box 278
LAKE ELSINORE, CA 92531-0278
Phone: 909-674-7041
Red Worms; Pit-run;
Castings, Potting Soil;
Mail order & Local
Wholesale & Retail

* * *

CONNECTICUT

James Plourde
Manchester Worm Farm
MANCHESTER,CT 06040
Phone: 203-647-8067
Nightcrawlers, Super Worms,
Rainbow Giant Meal Worms;
Fabrical Bait Cups;
Mail Order & Local
Wholesale & Retail
(See display ad pg. 46)

* * *

FLORIDA

Larry D. Martin
Vermi Technology Unlimited, Inc.
P.O. Box 130
ORANGE LAKE,FL 32681
Phone: 904-591-1111
Red Worms, Alabama Jumpers;
Mature, Bait, Pit-run, Capsules;
Castings, Vermi Composting bins,
Commercial bins, Consultants;
Mail Order & Local
Wholesale & Retail
(See display ad pg. 57)

* * *

Walter & Kay Smith
K&W Rabbit Farm
P.O. Box 50691
TICE, FL 33905-0691
Phone: 813-543-2620
Red Worms; Mature, Bait, Pit-run;
Worm Castings,
New Zealand White Rabbits;
Mail order & Local
Wholesale
(See display ad pg. 42)

* * *

J.A. Martin
The Bag Lady
18570 NE 43rd Ct. Rd.
CITRA, FL 32113
Phone: 904-591-1111
Breathable Earthworm Bags;
Mail Order & Local
Wholesale
(See display ad pg. 58)

* * *

A.L. (Larry) Williams
Steel City Worm Farm
2134 Hwy. 231 S.
COTTONDALE, FL 32431
Phone: 904-579-4985
904-579-4981
Red Worms, African Nightcrawlers,
Gray Wigglers (Florida Jumpers);
Mature, Bait Pit-run;
Bait cups; Mail Order & Local;
Wholesale
(See display ad pg. 31)

Charlie Morgan
P.O. Box 116G
BUSHNELL, FL 33513
Retired earthworm authority will answer questions directly related to a problem covered in one of his books. A stamped, self-addressed envelope is required. No S.A.S.E. no answer.

* * *

GEORGIA

Sybil Willingham
Willingham Worm Farm
Rt.1, Box 241
BUTLER, GA. 31006
Phone: 912-862-5545
Red Worms, Georgia Jumpers;
Pit-run; Mail Order; Wholesale
(See display ad pg. 55)

* * *

J. Frank Smith
Smith's Worms
Rt.2 Box 205
Boston, GA 31626
Phone: 912-498-1605
Red Worms; Mature, Pit-run;
Shipping Boxes, Bait Cups
Books, Instruction;
Mail Order & Local
Wholesale & Retail
(See display ad pg. 51)

* * *

David R. Stapleton
Stapleton Bait Farm
RFD 1
LUMBER CITY, GA 31549
Phone: 912-363-4451
Red Worms; Pit-run;
Mail Order, Wholesale
(See display ad pg. 36)

* * *

Jerry Seymour
Bronwood Worm Farms
P.O. Box 28
BRONWOOD, GA 31726-0028
Phone: 912-995-5994
Red Worms, Native Nightcrawlers;
Mature, Bait, Pit-run;
Mail order, Wholesale
(See display ad pg. 61)

* * *

IDAHO

Billy W. Dotson
Happy Hooker Fish Worms
B&J Enterprises
HC01, Box 17
PINEHURST, ID 83850
Phone: 208-682-4223
Red Worms, African Nightcrawlers,
Native Nightcrawlers;
Mature, Bait, Pit-run;
Mail Order & Local
Wholesale & Retail
(See display ad pg. 54)

ILLINOIS

Ray Edwards
The Bait Man
153 E. Woodlawn Ave.
P.O. Box 1233
HINCKLEY, IL 60520
Phone: 1-800-729-7312
Red Worms, Native Nightcrawlers;
Mature, Bait, Pit-run;
Water Dogs, Leeches, Worm Books,
Worm Bedding, Bait Vending Equipment;
Mail Order & Local
Wholesale & Retail
(See display ads pgs. 43 & 63)

INDIANA

Richard C. Beeson
Beeson's Worm Farm
1826 E. Elmore, Dept 1
CRAWFORDSVILLE, IN 47933
Phone: 1-317-362-0069
Red Worms; Mature, Bait;
Red Worm Eggs
Mail Order & Local
Wholesale & Retail

KANSAS

Lyle E. Dunham
L & T Enterprise Worm Ranch
119 N. 8th
HUMBOLDT, KS 66748
Phone: 316-473-2022
African Nightcrawlers;
Mature, Bait
Mail Order & local
Wholesale & Retail
(See display ad pg. 52)

KENTUCKY

Jack Leach
Leach's Bait Farm
Rt.1 Box 296-C
EAST BERNSTADT, KY 40729
Phone: 606-843-6597
Red Worms, African Nightcrawlers, Canadian Nightcrawlers, Grey Nightcrawlers,Jumpers, Giant Mealworms, Wax Worms, Golden Grubs, Crickets
Mail Order & Local
Wholesale & Retail
(See display ad pg. 27)

* * *

Don & Sherri DeArmond
DeArmond's Worm Farm
3923 Greenback Rd.
UTICA, KY 42376
Phone: 502-785-9137
Red Wigglers, Mature, Bait, Pit-run; Worm Bags, Shipping Boxes, Bags for Worm Castings;
Mail Order & Local
Wholesale & Retail
(See Display ad pg. 39)

* * *

Darrel & Linda Cravens
Cravens Worm Farm
Rt. 6, Box 174
LIBERTY, KY 42539
Phone: 606-787-8876
Red Worms; Mature, Bait, Pit-run; Worm Shipping Bags, Worm Castings;
Mail Order & Local
Wholesale & Retail

* * *

LOUISIANA

Marie Stewart
Stewart's Worm Farm
P.O. Box 188
SPRINGHILL, LA 71075
Phone: 318-847-4481
Red Worms; Pit-run; Castings;
Mail Order & Local
Wholesale
(See Display ad pg. 45)

* * *

Susan Damper
Southern Worm Enterprise
12118 Marilyn Ln.
HAMMOND, LA 70403
Phone: 504-294-2603
Red Worms, Brown Nose, Grey Nightcrawlers, Jumpers; Mature, Bait, Pit-run;
Mail Order & Local
Wholesale
(See display ad pg. 53)

* * *

Preston Welch
Monterey Worm Farm
9518 Hwy 129
MONTEREY, LA 71354
Phone: 318-386-2652
Red Worms; Pit-run;
Mail Order & Local
Wholesale & Retail
(See display ad pg. 46)

* * *

MICHIGAN

Michael Stockman
Superior Bait Farm
7701 Silver Lake Rd.
WOLVERINE, MI 49799
Phone: 616-525-8137
Red Worms, Native & Canadian Nightcrawlers, Wax Worms, Meal Worms, Mature, Bait, Pit-run; Crickets, Minnows; Bait cups, Wholesale Tackle, Wire cages;
Mail Order & Local
Wholesale & Retail
(See display ad pg. 37)

* * *

Lee R. Allen
Lee's Worm Ranch
2619 Bullock Rd.
BAY CITY, MI 48708
Phone: 517-893-4119
Red Worms; Mature, Bait, Pit-run; Castings;
Mail Order & Local
Wholesale & Retail
(See display ad pg. 56)

* * *

MISSISSIPPI

Howard or Marie Mays
Happy Chick Worm, Rabbit,
Cricket Ranch
14180 Mays Rd.
GULFPORT, MS 39503
Phone: 601-832-1270
Red Worms, African Nightcrawlers,
Grey Nightcrawlers;
Mature, Bait, Pit-run;
Cups, Worm Keeper, Rabbits &
Rabbit Supplies, Worm Feed, Comfrey
Mail Order & Local
Wholesale & Retail
(See display ad on back cover)

* * *

John F. Nicholson
Nicholson's Worms
873 Green Acres
HERNANDO, MS 38632
Phone: 601-429-5171
Red Worms,; Mature, Bait,
Pit-run, Capsules; Books,
Bedding, consultation;
Mail Order & Local
Wholesale & Retail

Richard & Shirlene Phelps
Phelps Worm Farm
Rt.1 , Box 433
SILVA, MO 63964
Phone: 314-783-3944
Red Worms, Native Nightcrawlers;
Mature, Bait, Pit-run;
Books, cloth shipping bags;
Mail Order & Local
Wholesale & Retail
(See display ad pg. 52)

* * *

Wally Kiger
Wally's Wigglers Worm & Ginseng Farm
Rt.1, Box 107
STOUTSVILLE, MO 65283
Phone: 314-672-3233
Red Worms, Tiger Worm;
Mature, Bait, Pit-run;
Mail order & Local
Wholesale

* * *

MISSOURI

Dave Metzger
Earthworms & Supplies
Rt.3, Box 291
OWENSVILLE, MO 65066
Phone: 314-437-3605
Mature Worms;
Bags, Feed Grinder, Worm Separator;
Mail order & Local
Wholesale & Retail
(See display ad pg. 60)

* * *

Douglas Swan
Swan's Bait Farm
Rt.1, Box 244
RAYMONDVILLE, MO 65555
Phone: 417-457-6215
Giant Native Worms;
Mature, Bait, Pit-run;
Mail Order & Local
Wholesale
(See display ad pg. 50)

Joberta Crossin
Granny's Wigglers
P.O. Box 186
CLARKTON, MO 63837
Phone: 314-448-3302
Red Worms; Pit-run;
Mail order & Local
Wholesale & Retail

* * *

Marvin A. Dickman
Ozark Worm Farm
Rt.3, Box 160-D
WILLOW SPRINGS, MO 65793
Phone: 417-469-2286
Treated Wood Worm Beds,
Ginseng & Earthworm Books;
Mail Order; Wholesale & Retail
Registered Cairn Terriers
& Persian cats.

* * *

Bill Beck
Lucky Worm Farm
1807 Venda
KENNET, MO 63857
Phone: 314-888-0705
Red Worms, Nightcrawlers, Jumpers;
Books, cups, free advice;
Mail order & Local
Wholesale & Retail

* * *

Bob Thomson & Ivan Broyles
Brookside Worm Farm
P.O. Box 386
GOWANDA, NY 14070-0386
Phone: 716-532-2999 or 3483
Red Worms, Jumpers,
Brown Nose, Capsules;
Year 'round mail order,
Local, export
Wholesale & Retail

* * *

NEBRASKA

Chuck Hutter
H & K Research & Development Co.,Inc.
P.O. Box 44184
OMAHA, NE 68144
Phone: 402-333-5669
Worm Bed Plans
(See Display ad pg. 47)

* * *

Fred Patz
Fred's Worm Ranch
1409 N. 60th
LINCOLN, NE 68505
Phone: 402-464-0994
Red Worms, African Nightcrawlers,
Native Nightcrawlers;
Mature, Bait, Pit-run;
Worm publications, Starter kit, Feed;
Mail order; Wholesale & Retail
(See Display ad pg. 48)

* * *

NEW YORK

Gene Artz
Sealright Packaging Company
314 South First St.
FULTON, NY 13069
Phone: 315-593-5311
Sealright Supplies, Moisture-resistant containers for fish bait and gardening industries;
Distributors & Direct , wholesale
(See display ad pg. 35)

NORTH CAROLINA

Gene Shepherd
Foothills Bait Company & Rabbitry
Rt.1, Box936
NORTH WILKESBORO, NC 28659
Phone: 910-670-9360
1-800-487-7878
FAX: 910-670-9425
Red Wigglers, Brown Nose,
African Nightcrawlers;
Mature, Bait, Pit-run
Mail Order & Local, Wholesale
(See display ad pg. 49)

* * *

Prentis & Annie Campbell
Campbell Worm Farm
207-A Ramsey Dr.
JACKSONVILLE, NC 28540
Phone: 910-346-4565
Red Worms; Bait, Pit-run;
Spikes (Gray larva Maggots)
Worm Food, Potting Soil,
Worm Bags (for shipping)
Mail Order & Local,
Wholesale & retail
(See display ad pg. 48)

* * *

George V. Parker
Boot's Worm Farm
Rt. 7, Box 575-A
LUMBERTON, N.C. 28358
Phone: 910-738-7432
Red Worms; Pit-run;
Mail Order & Local
Wholesale & Retail
(See Display ad pg. 61)

James A. Cannon
Cannon's Earthworms
Rt. 12, Box 877
HICKORY, NC 28602
Phone: 704-294-3249
Red Worms, Native Nightcrawlers;
Mature, Bait, Pit-run;
Worm food, Worm books,
cloth shipping bags;
Wholesale & Retail
Mail order & Local

* * *

OHIO

Connie, John Focht
Worms A-Lure
194 Ansel Road
GENEVA, OH 44041
Phone: 216-466-4797
Red Worms; Bed run, Breeders;
Bedding, Worm bins
Mail Order & Local
Wholesale & Retail
(See display ad pg. 36)

* * *

Carol Miller
C & G Enterprises
P.O. Box 658
PERRY, OH 44081
Phone: 216-354-5414
Red Worms; Mature, Bait;
Mail Order & Local
Wholesale & Retail
(See display ad pg. 51)

* * *

Gary & Don Wolcott
Regional Bait & Marketing
8006 Plank Road
THOMASON, OH 44086
Phone: 216-298-9766
1-800-RED WORM
Red Worms; Mature, Bait, Pit-run;
Capsules; Mail Order & Local
Wholesale & Retail
(See display ad pg. 62)

OKLAHOMA

Rosemary Brown
Brown's Worm Farm
Box 284
MARIETTA, OK 73448
Phone: 405-276-3897
Brown Nose worms; Pit-run;
Mail Order, Wholesale

* * *

PENNSYLVANIA

Kay Ihnat
Worm World
RD 2, Box 188, Ihnat Lane
AVELLA, PA 15312
Phone: 412-356-2397
Red Worms; Mature, Bait, Pit-run;
Mail Order & Local
Wholesale & Retail
(See display ad pg. 61)

* * *

Tom & Betty Kroh
Woehrle's Wholesale Bait & Supplies,Inc.
95 Church Ave.
MT. POCONO, PA 18344
Phone: 717-839-8952
Red Worms, Native Nightcrawlers; Bait;
Live Bait supplies;
Mail Order & Local
Wholesale
(See display ad pg. 60)

* * *

Donald Hogan
Hogan's
RD 1, Box 1032
CLARENDON, PA 16313
Phone: 814-723-8417
Red Worms; Mature, Bait, Pit-run;
Earthworm Books;
Mail Order & Local;
Wholesale & Retail
(See display ad pg. 36)

* * *

SOUTH CAROLINA

R.J. "Butch" Melton
Melton Bait Farm
P.O. Box 895
MANNING, SC 29102
Phone: 803-478-4585
Red Worms, African Nightcrawlers;
Mature, Bait, Pit-run;
Mini Farms; Mail Order & Local;
Wholesale & Retail
(See display ad pg. 46)

* * *

Isaac L. McGill
Mac's Redworm Farm
108 King Ave.
KINGSTREE,SC 29556
Phone: 803-382-8119
Red Worms; Mature, Pit-run:
Mail Order & Local
Retail
(See display ad pg. 44)

* * *

Barbara E. Corson
Carolina Worm Farm
209 Calhoun St.
CLOVER, SC 29710
Phone: 803-222-3603
Red Worms; Mature, Bait, Pit-run;
Mail Order & Local
Wholesale

* * *

TENNESSEE

Otis Holcomb
Cross Road Baits
5375 Hwy. 321 N.
LENOIR CITY, TN 37771
Phone: 615-988-3853
African Worms; Bait;
Plastic Bait Cups
Mail Order & Local
Wholesale
(See display ad pg. 56)

Tracy Laster
Jet Worm Farm
P.O. Box 7736
JACKSON, TN 38308
Phone: 901-424-0154
Red Worms; Bait, Pit-run;
Mail order & local
Wholesale & Retail

* * *

Jack McCathran
Pleasant Shade Worm Ranch
Rt. 1, Box 47 BB
PLEASANT SHADE, TN 37145
Phone: 615-374-4901
Red Worms, Native Nightcrawlers;
Bait, Pit-run;
Mail Order & Local
Wholesale & Retail

* * *

TEXAS

Oscar & Irene Fincher
Oscar's Bait Farm
Rt.3, Box 49X
STEPHENVILLE, TX 76401
Phone: 817-968-5472
Red Worms, Tennessee Brown Nose;
Pit-run; Bait Containers,
Worm Signs, Worm Feed;
Mail Order & Local
Wholesale & Retail
(See display ad pg. 25)

* * *

J.A. "Jay" Mertz
Rabbit Hill Farm
Rt.3, Box 2936
CORSICANA, TX 75110
Phone: 903-872-4289
Brown Nose worms; Pit-run;
Mail Order & Local
Wholesale & Retail
(See display ad pg. 54)

* * *

Forrest Lemley
Lemley Worm Farm
P.O. Box 534
CEDAR HILL, TX 75104
Phone: 214-291-4632
Tennessee Brown Nose Worms;
Bait Size; Mail Order & Local

* * *

UTAH

John L. & Velma E. Ward
Lucky Hooker Worm Farm
295 N. State St.
FERRON, UT 84523
Phone: 801-384-2904
Red Worms; Pit-run;
Mail Order & Local
Wholesale & Retail
(See display ad pg. 58)

* * *

Arland Malmstrom
Mel's Worm Farm
450 S. 100 East
PRICE, UT 84501
Phone: 801-637-6910
Red Worms; Mail Order & Local;
Wholesale & Retail

* * *

Verlo Ottley
2186 Kirkham Way
SALT LAKE CITY, UT 84119
Phone: 801-968-5483
Red Worms, Native Nightcrawlers;
Pro worm cupping material,
"Fishworm" signs (8½ x 14)
I buy bulk nightcrawlers
year 'round, Local;
Wholesale & retail (send SASE)

* * *

VIRGINIA

Mike Nixon
Nixon & sons Live Baits (Worms 4U)
Rt. 2, Box 63
EVINGTON, VA 24550
Phone: 804-821-3191
Red Worms, Native & Canadian Night-
crawlers, Alabama Jumpers,
Mealworms, Waxworms; Mature, Bait;
Mail Order & Local
Wholesale & Retail
(See display ad pg. 54)

* * *

WISCONSIN

Carroll Abendroth
Bait Barn Worm Farm
W8697 Island Road
WATERLOO, WI 53594
Phone: 414-478-2053
Red Worms, Native Nightcrawlers;
Mature, Pit-run;
Books, cups, Misc. supplies,
Vending machines, worm harvesters;
Mail order & Local
Wholesale & retail
(See display ads pgs. 40 & 41)

* * *

Lori Sharkey
Sharkey's Worms
530 S.E. Baker Ave.
GRESHAM, WI 97080
Phone: 503-667-7826
Native nightcrawlers; Mature worms;
Wholesale

* * *

CANADA

Anna Scott Draper
Early Bird Ecology & Bait Farms Ltd.
RR 1
SMITHVILLE, ONT. LOR 2AO
Canada
Phone: 416-643-4252
Red Worms, African Nightcrawlers;
Mature, Bait; Dehydrated worms,
Worm Logo Apparel, Worm Books,
Kitchen Komposters;
Mail order & Local;
Wholesale & Retail
(See display ad pgs. 32 & 33)

* * *

Janet & Jack Livingstone
The Worm Factory
129 Sherbrooke St. E.
PERTH, ONT. K7H 1B1
Canada
Red Worms; Pit-run; Books,
Bedding, Indoor composting kits;
Mail order & local;
Wholesale & retail
(See display ad pg. 53)

* * *

Wendy MacDonell
Goldwen Ecosystems
RR 1
PARIS, ONT. N3L 3E1
Canada
Phone: 519-756-6469
Red Worms;, Mature, Bait, Pit-run;
Mulched paper, worm bedding;
Mail order & Local;
Wholesale
(See display ad pg. 55)

* * *

Paul L. Fretz
Paul Fretz Ent. Inc.
Winterbourne, RR 2, Rd. 23
WEST MONTROSE, ONT. NOB 2VO
Canada
Phone: 519-664-2701
Red Worms; Pit-run;
Mail order, Retail
(See display ad pg. 62)

* * *

Cindy & Phil
Hermi The Wormi Vermiculture Corporation
207-13931-74 Ave.
SURREY, BC V3W-6G6 , Canada
Phone: 604-572-1433
Red Worms; Bait, Pit-run;
Castings, calcium carbonate
complete indoor worm bins("Hermi's
Kitchen Condos") ;Mail order & Local
Wholesale & Retail

* * *

John & Glenna Armstrong
Down To Earth-worms, Composting
& Bait Supplies
2249 Valleyview Dr.
KAMLOOPS, B.C. V2C 4C8 ,
Canada
Phone: 604-374-5718
Red Worms; Mature, Bait, Pit-run;
Worm Mate Vermicomposters,
Harvesters, Shipping Bags;
Mail Order & Local
Wholesale & Retail

* * *

Michael Scholey
Mike's Bait & Tackle
546 Dupplin Road
VICTORIA, B.C. V8Z 1C1
Canada
Phone: 604-383-9081
Native Nightcrawlers; Bait;
Salmon Eggs Bait, Fishing Tackle,
Worm Castings; Mail order & Local;
Wholesale

* * *

CLASSIFIED LISTINGS

(See General Listings for complete addresses)

RED WORMS

STATE	CITY	NAME
ALABAMA	Tuskegee Institute	Hall's Organic Gardens
	Chatom	Tate's Worm & Fern Farm
ARKANSAS	Yellville	Worms By Weber
	Benton	Griffin Farms
	Wiseman	Powell's Bait Farm
	Paragould	Family Worm Farm
	Paragould	J&S Worm Farms
	Rogers	National Rabbit Enterprises
	Batesville	John's Worm Ranch
	Paris	Joan Hartfiel
	Paragould	Filerbo Worm Farm
	Ft. Smith	Red-Eye Bait Farm
	Siloam Springs	Wholesale Worms
CALIFORNIA	Granada Hills	Hy Hunter Industries
	Davis	Rainbow Worm Farm
	Lincoln	Wormco
	Dixon	Cain's Farm
	Vacaville	Denton's Quality Worm Farm
	Simi Valley	The Worm Concern
	Blue Lake	Redwood Valley Worm Ranch
	Pomona	Aarion Wormery I
	Yucca Valley	Kurt E. Papenfus
	Lake Elisnore	The Potted Thorne Nursery
	Suisun	Earthworks Enterprises
FLORIDA	Orange Lake	Vermi Technology Unlimited,Inc.
	Tice	K&W Rabbit Farm
	Cottondale	Steel City Worm Farm
GEORGIA	Butler	Willingham Worm Farm
	Boston	Smith's Worms
	Lumber City	Stapleton Worm Farm
	Bronwood	Bronwood Worm Farms
IDAHO	Pinehurst	Happy Hooker Fish Worms
ILLINOIS	Hinckley	The Bait Man
INDIANA	Crawfordsville	Beeson's Worm Farm
KENTUCKY	East Bernstadt	Leach's Bait Farm
	Utica	DeArmond's Worm Farm
	Liberty	Cravens Worm Farm
LOUISIANA	Springhill	Stewart's Worm Farm
	Hammond	Southern Worm Enterprise
	Monterey	Monterey Worm Farm
MICHIGAN	Wolverine	Superior Bait Farm

RED WORMS CONT.

MISSISSIPPI	Gulfport	Happy Chick Worm,Rabbit,Cricket Farm
	Hernando	Nicholson's Worms
MISSOURI	Silva	Phelps Worm Farm
	Stoutsville	Wally's Wigglers Worm & Ginseng Farm
	Clarkton	Granny's Wigglers
	Kennet	Lucky Worm Farm
NEBRASKA	Lincoln	Fred's Worm Ranch
NEW YORK	Gowanda	Brookside Worm Farm
NORTH CAROLINA	N. Wilkesboro	Foothills Bait Co.& Rqbbitry
	Jacksonville	Campbell Worm Farm
	Lumberton	Boots Worm Farm
	Hickory	Cannon's Earthworms
OHIO	Geneva	Worms A-Lure
	Perry	C&G Enterprises
	Thomason	Regional Bait & Marketing
PENNSYLVANIA	Avella	Worm World
	Mt. Pocono	Woehrle's Wholesale Bait
	Clarendon	Hogan's
SOUTH CAROLINA	Manning	Melton Bait Farm
	Kingstree	Mac's Redworm Farm
	Clover	Carolina Worm Farm
TENNESSEE	Jackson	Jet Worm Farm
	Pleasant Shade	Pleasant Shade Worm Ranch
TEXAS	Stephenville	Oscar's Bait Farm
UTAH	Ferron	Lucky Hooker Worm Farm
	Price	Mel's Worm Farm
	Salt Lake City	Verlo Ottley
VIRGINIA	Evington	Nixon & Sons Live Baits
WISCONSIN	Waterloo	Bait Barn Worm Farm
CANADA	Smithville, ONT.	Early Bird Ecology & Bait Farms Ltd.
	Perth, ONT.	The Worm Fqctory
	Paris, ONT.	Goldwen Ecosystems
	W. Montrose, ONT.	Paul Fretz Ent., Inc.
	Surrey, B.C.	Hermi The Wormi Vermiculture Corp.
	Kamloops, B.C.	Down to Earth-Worms, Composting & Bait Supplies

AFRICAN NIGHTCRAWLERS

ARKANSAS	Yellville	Worms By Weber
	Benton	Griffin Farms
CALIFORNIA	Granada Hills	Hy Hunter Industries

(African Nightcrawlers Cont.)

FLORIDA	Cottondale	Steel City Worm Farm
IDAHO	Pinehurst	Happy Hooker Fish Worms
KANSAS	Humboldt	L&T Enterprise Worm Ranch
KENTUCKY	East Bernstadt	Leach's Bait Farm
MISSISSIPPI	Gulfport	Happy Chick Worm,Rabbit,Cricket Ranch
NEBRASKA	Lincoln	Fred's Worm Ranch
NORTH CAROLINA	N. Wilkesboro	Foothills Bait Co.& Rabbitry
SOUTH CAROLINA	Manning	Melton Bait Farm
TENNESSEE	Lenoir City	Cross Road Baits
WISCONSIN	Waterloo	Bait Barn Worm Farm
CANADA	Smithville, Ont.	Early Bird Ecology & Bait Farms Ltd.

NATIVE NIGHTCRAWLERS

ARKANSAS	Ft. Smith	Red-Eye Bait Farm
CALIFORNIA	Granada Hills Simi Valley	Hy Hunter Industries The Worm Concern
CONNECTICUT	Manchester	Manchester Worm Farm
GEORGIA	Bronwood	Bronwood Worm Farms
IDAHO	Pinehurst	Happy Hooker Fish Worms
ILLINOIS	Hinckley	The Bait Man
KENTUCKY	East Bernstadt	Leach's Bait Farm
MICHIGAN	Wolverine	Superior Bait Farm
MISSOURI	Silva Kennet	Phelps Worm Farm Lucky Worm Farm
NEBRASKA	Lincoln	Fred's Worm Ranch
NORTH CAROLINA	Hickory	Cannon's Earthworms
PENNSYLVANIA	Mt. Pocono	Woehrle's Wholesale Bait
TENNESSEE	Pleasant Shade	Pleasant Shade Worm Ranch
UTAH	Salt Lake City	Verlo Ottley
VIRGINIA	Evington	Nixon & Sons Live Baits
WISCONSIN	Waterloo	Bait Barn Worm Farm
CANADA	Victoria,B.C.	Mike's Bait & Tackle

GREY NIGHTCRAWLERS

ALABAMA	Chatom	Tate's Worm & Fern Farm
CALIFORNIA	Vacaville	Denton's Quality Worm Farm
FLORIDA	Cottondale	Steel City Worm Farm
KENTUCKY	East Bernstadt	Leach's Bait Farm
LOUISIANA	Hammond	Southern Worm Enterprise
MISSISSIPPI	Gulfport	Happy Chick Worm,Rabbit,Cricket Ranch

JUMPERS

CALIFORNIA	Vacaville	Denton's Quality Worm Farm
FLORIDA	Orange Lake Cottondale	Vermi Technology Unlimited,Inc. Steel City Worm Farm
GEORGIA	Butler	Willingham Worm Farm
KENTUCKY	East Bernstadt	Leach's Bait Farm
LOUISIANA	Hammond	Southern Worm Enterprise
MISSOURI	Kennet	Lucky Worm Farm
NEW YORK	Gowanda	Brookside Worm Farm
VIRGINIA	Evington	Nixon & Sons Live Baits

BROWN NOSE WORMS

ALABAMA	Chatom	Tate's Worm & Fern Farm
ARKANSAS	Eudora	Jim Knight
CALIFORNIA	Granada Hills	Hy Hunter Industries
LOUISIANA	Hammond	Southern Worm Enterprises
NEW YORK	Gowanda	Brookside Worm Farm
NORTH CAROLINA	N. Wilkesboro	Foothills Bait Co. & Rabbitry
OKLAHOMA	Marietta	Brown's Worm Farm
TEXAS	Stephenville Corsicana Cedar Hill	Oscar's Bait Farm Rabbit Hill Farm Lemley Worm Farm

OTHER TYPES OF BAIT

Catfish Bait Worms

Catfish Bait Worms	National Rabbit Enterprises, Rogers, AR
Crickets	Rainbow Mealworms, Compton, CA Kurt E. Papenfuhs, Yucca Valley, CA Leach's Bait Farm, East Bernstadt, KY Superior Bait Farm, Wolverine, MI
Golden Grubs	Leach's Bait Farm. East Bernstadt, KY
Leeches	The Bait Man, Hinckley, IL
Meal Worms	Rainbow Mealworms, Compton, CA Wormco, Lincoln, CA Manchester Worm Farm, Manchester, CT Leach's Bait Farm, E. Bernstadt, KY Superior Bait Farm, Wolverine, MI Nixon & Sons Live Baits, Evington, VA
Salmon Eggs	Mike's Bait & Tackle, Victoria, B.C., Canada
Spikes (Maggots)	Campbell Worm Farm, Jacksonville,NC
Minnows	Superior Bait Farm, Wolverine, MI
Tiger Worms	Worms by Weber, Yellville, AR Griffin Farms, Benton, AR Wally's Wigglers Worm & Ginseng Farm, Stoutsville, MO
Water Dogs	The Bait Man, Hinckley, IL
Wax Worms	Rainbow Mealworms,Inc., Compton, CA Leach's Bait Farm, E. Bernstadt, KY Superior Bait Farm, Wolverine, MI Nixon & Sons Live Baits, Evington, VA

MISCELLANEOUS

Bait Containers Shipping & Packaging Supplies	Jim Knight, Eudora, AR J&S Worm Farms, Pargould, AR Rainbow Worm Farm, Davis, CA Manchester Worm Farm, Manchester, CT The Bag Lady, Citra, FL Steel City Worm Farm, Cottondale, FL Smith's Worms, Boston, GA DeArmond's Worm Farm, Utica, KY Cravens Worm Farm, Liberty, KY Superior Bait Farm, Wolverine, MI Happy Chick Worm,Rabbit, Cricket Ranch, Gulfport, MS Earthworms & Supplies, Owensville, MO Phelps Worm Farm, Silva, MO Sealright Packaging Co., Fulton, NY

	Campbell Worm Farm, Jacksonville, NC Cannon's Earthworms, Hickory, NC Cross Road Baits, Lenoir City, TN Oscar's Bait Farm, Stephenville, TX Verlo Ottley, Salt Lake City, UT Bait Barn Worm Farm, Waterloo, WI Down To Earth-Worms, Composting & Bait Supplies, Kamloops, B.C. Canada
Bait Vending Equipment	The Bait Man, Hinckley, IL Bait Barn Worm Farm, Waterloo, WI
Calcium Carbonate	Hermi The Wormi Vermiculture Corp.,Surrey,B.C. Canada
Comfrey	Happy Chick Worm,Rabbit,Cricket Ranch, Gulfport, MS
Compost	Hall's Organic Gardens, Tuskegee Institute, AL
Composting Bins & Kits	Vermitechnology Unlimited,Inc., Orange Lake,FL The Worm Factory, Perth, Ont., Canada Down To Earth-Worms Composting & Bait Supplies, Kamloops, B.C. Canada
Consultants	Vermitechnology Unlimited, Inc.,Orange Lake, FL Lucky Worm Farm, Kennet, MO
Dehydrated Worms	Early Bird Ecology & Bait Farms Ltd., Smithville, Ont. CAN
Earthworm Books, Instructions	National Rabbit Enterprises, Rogers, AR John's Worm Ranch, Batesville, AR Joan Hartfiel, Paris, AR Wholesale Worms, Siloam Springs, AR Smith's Worms, Boston, GA The Bait Man, Hinckley, IL Nicholson's Worms, Hernando, MS Phelps Worm Farm, Silva, MO Ozark Worm Farm, Willow Springs, MO Fred's Worm Ranch, Lincoln, NE Cannon's Earthworms, Hickory, NC Hogan's, Clarendon, PA Bait Barn Worm Farm, Waterloo, WI Early Bird Ecology & Bait Farms,Ltd.,Smithville,Ont. CAN. The Worm Factory, Perth, Ont., Canada
Fishing Tackle	Superior Bait Farm, Wolverine, MI Mike's Bait & Tackle, Victoria, B.C.,Canada
Kitchen Komposters	Early Bird Ecology & Bait Farms,Ltd.,Smithville,Ont. CAN.
Live Bait Supplies	Woehrle's Wholesale Bait & Supplies,Inc., Mt. Pocono,PA
Manure	Wormco, Lincoln, CA
Mini Farms	Melton Bait Farm, Manning, SC
Organically Grown Plants	Hall's Organic Gardens, Tuskegee Institute, AL The Worm Concern, Simi Valley, CA
pH Testers	Joan Hartfiel, Paris, AR
Organic Garden & Lawn Supplies	Worms by Weber, Yellville, AR
Potting & Starter Soil	Hall's Organic Gardens, Tuskegee Institute, AL Rainbow Worm Farm, Davis, CA Wormco, Lincoln, CA

(Miscellaneous Cont.)

	The Worm Concern, Simi Valley, CA The Potted Thorne Nursery, Lake Elsinore, CA Campbell Worm Farm, Jacksonville, NC
Rabbits & Rabbit Supplies	National Rabbit Enterprises, Rogers, AR K&W Rabbit Farm, Tice, FL Happy Chick Worm,Rabbit,Cricket Ranch, Gulfport,MS
Signs (worm)	Oscar's Bait Farm, Stephenville, TX Verlo Ottley, Salt Lake City, UT
Starter Kits	Kurt E. Papenfuhs, Yucca Valley, CA Fred's Worm Ranch, Lincoln, NE
Suppies (General)	National Rabbit Enterprises, Rogers, AR Hy hunter industries, Granada Hills, CA Bait Barn Worm Farm, Waterloo, WI
Tumblers	Jim Knight, Eudora, AR
Video Instructions	Jim Knight, Eudora, AR Norb's Bait Farm, Norco, CA
Wire Cages	Mike's Bait & Tackle, Victoria, B.C., CAN.
Worm Bedding	The Bait Man, Hinckley, IL Nicholson's Worms, Hernando, MS Worms-A-lure, Geneva, OH The Worm Factory, Perth, Ont., CAN. Goldwen Ecosystems, Paris, Ont., CAN.
Worm Bed Plans	H&K Research & Development Co.,Omaha, NE
Worm Beds & Bins	Vermi Technology Unlimited,Inc.,Orange Lake,,FL Ozark Worm Farm, Willow Springs, MO Worms-A-Lure, Geneva, OH Hermi The Wormi Vermiculture Corp.,Surrey,B.C.,CAN
Worm Castings	Worms by Weber, Yellville, AR Griffin Farms, Benton, AR Rainbow Worm Farm, Davis, CA Wormco, Lincoln, CA Cain's Farm, Dixon, CA Denton's Quality Worms, Vacaville, CA The Worm Concern, Simi Valley, CA Redwood Valley Worm Farm,Blue Lake, CA The Potted Thorne Nursery, Lake Elsinore, CA Earthworks Enterprises, Suisun, CA Vermi Technology Unlimited,Inc., Orange Lake,FL K&W Rabbit Farm, Tice, FL Cravens Worm Farm, Liberty,KY Stewart's Worm Farm,. Springhill, LA Lee's Worm Ranch, Bay City, MI Nicholson's Worms,. Hernando, MS Hermi The Wormi Vermiculture Corp.,Surrey,B.C. CAN. Mike's Bait & Tackle, Victoria, B.C., CAN.
Worm Capsules	Beeson's Worm Farm, Crawfordsville, IN Nicholson's Worms, Hernando, MS Brookside Worm Farm, Gowanda, NY Regional Bait & Marketing, Thomason, OH

Worm Feed	Happy Chick Worm,Rabbit,Cricket Ranch, Gulfport, MS Fred's Worm Ranch, Lincoln, NE Campbell Worm Farm, Jacksonville, NC Cannon's Earthworms, Hickory, NC Oscar's Bait Farm, Stephenville, TX
Worm Feed Grinder	Earthworms & Supplies, Owensville, MO
Worm Harvesters	Joan Hartfiel, Paris, AR Rainbow Worm Farm, Davis, CA Earthworms & Supplies, Owensville, MO Bait Barn Worm Farm, Waterloo, WI Down To Earth-Worms, Composting & Bait Supplies, Kamloops, B.C. Canada
Worm Keeper	Happy Chick Worm, Rabbit, Cricket Ranch, Gulfport,MS
Worm School	Worms by Weber, Yellville, AR
Worm Logo Apparel	Early Bired Ecology & Bait Farms,Ltd., Smithville,Ont.,CAN

INDEX TO ADVERTISERS

NOW YOU HAVE A CHOICE OF BAIT CONTAINERS

Heavy Duty economical 8 oz., 12 oz., and 15 oz. styrofoam bait cups with 12 vent grooves around the styrofoam lid that plugs into the cup for secure closing. Keeps worms in better condition for a longer period.

Sample of 25 cups & lids $6.00 postpaid

We can also supply you with:

- Bedrun Red Worms
- Bedrun Tennessee Brown Nose Worms
- Worm Castings
- Worm Signs
- Worm Publications
- Worm Feed

Write or call for free information and prices

Oscar's Bait Farm

Route 3, Box 49X-1
Stephenville, Texas 76401
Phone 817/968-5472

Moore's
Rabbitry & Worm Farm

We carry:

- Bedrun Red Worms
- By the pound, cup, or ton

3 Lb. = $24.00
5 Lb. = $37.50
10 Lb. or more $6.50

All Prices Postpaid

Free Delivery on orders of 250 lbs. or more within 500 miles.

Available also:

- New Zealand & California Rabbits
- Rabbit Watering System
- Custom Built Rabbit Cages

Soon to come:

- Brown Crickets

Moore's
Rabbitry & Worm Farm
P.O. Box 471
Opelika, AL 36801
Phone: 205-705-0035

Since 1979

Specializing in African Nightcrawlers

Red Wigglers
African Nightcrawlers
Canadian Nightcrawlers
Grey Nightcrawlers,
Jumpers, Wigglers

Crickets
Giant Mealworms
Wax Worms
Golden Grubs
Books on Growing Worms

Cups or Bulk

We will work with you on any size order. Freight is paid on any minimum size order that can be shipped by surface UPS. Shipping time and temperature are factors to consider. Prices subject to change.

Jack Leach

LEACH'S BAIT FARM®

333 Rocky Branch Rd • East Bernstadt, Kentucky 40729
Tel.: 606/843-6597

(IMPORTANT: When writing, be sure to give us your phone number. Because my time is limited, I will answer mail requests by phone.)

Hy Hunter The WormKing buys and sells Millions and Millions of Worms

IN THE EARTHWORM BUSINESS SINCE 1931

Dealers, Jobbers and Distributors needed

Buy — Sell - Trade/Wholesale & Retail

Listed here are a few of the items we have in stock for immediate delivery:

Red Worms
African Nightcrawlers
Brown Nose Worms
Native Nightcrawlers
Gray Nightcrawlers
Garden & Dug Worms
California Tiger Worms
Calif. Golden Giant Superworms (T.M.)
Hawaiian Dancer Worms (T.M.)
Hulu Worms (T.M.)
Wonder Reds (T.M.)
Superworms (T.M.)
Wharfworms (T.M.) (Best for ocean fishing)
Bait Containers
Bedding
Bins, Superworm Farms
Bin Sales
Earthworm Books
Cups
Feeds, Special Earthworm Mix Which Includes Vitamins
Heating Tapes for Worm Bins
Lime
Manure
Organic Fertilizer "Wormilizer" (T.M.)
Peat Moss, California, Canadian, Florida
Rabbit Books
Potting Mix
Soil Plant Food
Super Giant Mulcher (T.M.) Mulches Two Hundred Tons an Hour
Green Belt (T.M.) A New Drip Irrigation Water System
Rex Chinchillas
Worm Guard, Keeps Worms In Their Bins
Supplies
Vending Machines for Worms
Worm Bed Plans ($5.00)
Worm Castings
Worm Culture
Worm Packing
Worm Shipping Cartons
Rex Superrabbits
Rabbit Cages
California Green Giant Frogs (Some of the world's largest)
Araucanas Chickens (Lay Colored Easter Eggs Year Around)
King White Turkeys (Superturkeys, weight over 100 lbs.)
Ducks
Geese
Chickens, Plain & Fancy Breeds
Game Birds
Coturnix Quail
Catfish, Silver & Gold
Trout
Bass
Shrimp
Lobster, Eastern & Western
Mini Sea Aquaculture Systems
Fish Rearing Loop Systems
Aquatic Loop Fish Production Systems
Fish & Frog Rearing Tanks

We are interested in opening branch offices throughout the U.S.A. and foreign countries. If you would be interested in further information, please write.

Hy Hunter the WormKing

Now Over 62 Years in the Earthworm Business

Hy Hunter Industries Inc.
Environmental & Agricultural Systems

Main Office and Home Ranch:
12001 Highwater Road
Granada Hills, California 91344
Phone (818) 363-0202

Can supply over 200 pages of the exciting *Whole Earthworm Catalogue #One*

ONLY $30.00 POSTPAID

Gentlemen Farmers . . . SENSATIONAL SYSTEM

NEW HYDRO GREENHOUSES

ALMOST UNBELIEVABLE PRODUCTION.

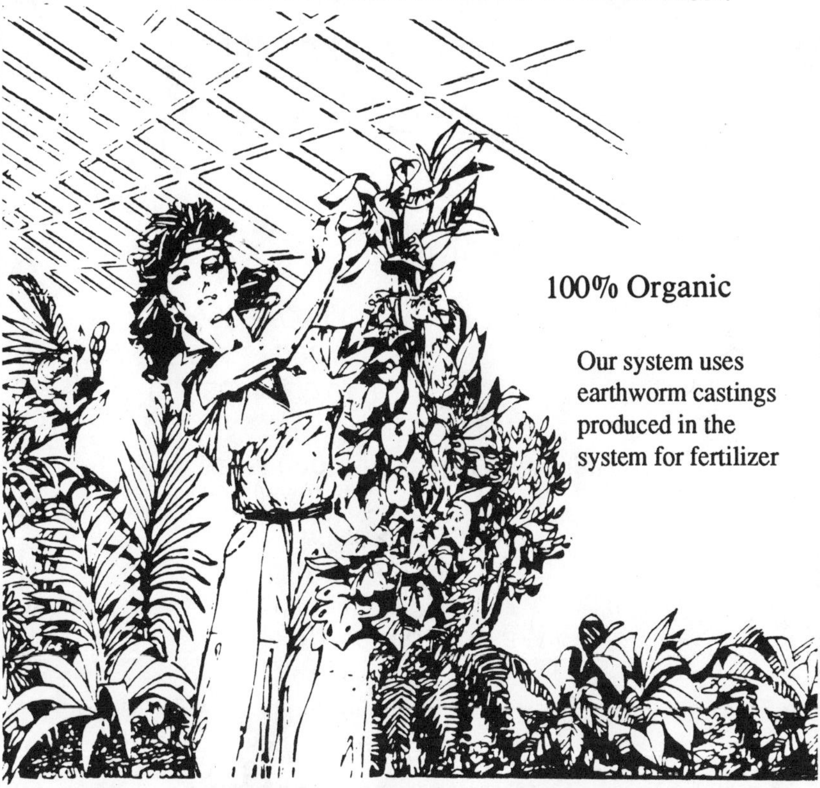

A NEW HYDROPONIC METHOD of raising flowers, vegetables, plants, trees, **earthworms,** fish, and possibly freshwater lobsters that can grow up to five pounds each. Our new method can produce up to ten times more product than a normal garden production of the same size. We can supply all size greenhouses from a small patio to huge commercial opeartions with complete components to put you in immediate production. Grow for yourself or supply supermarkets for **BIG PROFITS.**

GIANT FRESH WATER LOBSTER

GROWERS MANUAL Only **$100.00** Postpaid.

To receive brochures on Hydro Greenhouses, send $20.00 to:

HYDRO GREENHOUSE CORP.

12001 Highwater Road, Building 275

Granada Hills, California 91344 • (818) 363-0202

Rainbow Mealworms

Golden Mealworms:

Fishing size:	1,000 –	$5.00
	3,000 –	$13.50
	10,000 –	$35.00
Rainbow Giants:	1,000 –	$9.00

plus postage

We now have crickets and wax worms

We are looking for Distributors and Jobbers!

P.O. Box 4907
Compton, CA 90224

1-800-777-WORM

Mealworms are *excellent* for ice fishing!

Make Big $$$
In Earthworm Production

New professional ***Video*** shows the viewer exactly how to go about building a profitable business. They are led step by step so that they master all of the skills. Once they have their Earthworm business in production, they will find it easy to sell their output at big profits. Those who learn to raise earthworms can also save money by using them for gardening, soil enrichment, and fishing!!

Learn at home with our easy to understand ***Video!!***

Satisfaction Guaranteed • $34.95 Postpaid

Norb's Bait Farm • 1370 2nd St. • Norco, CA 91760

We have 25 acres of Wigglers
or Florida Jumpers...our best seller!
Also some Red Wigglers & African Nightcrawlers

We sell one million per week during summer season.

Steel City Worm Farm

Located 2 miles south of I 10 on U.S. 231
45 miles north of Panama City, Florida

2134 Hwy. 231 S.
Cottondale, Florida 32431

A.L. (Larry) Williams

Phone 904-579-4985
FAX 904-579-4127

WELCOME TO

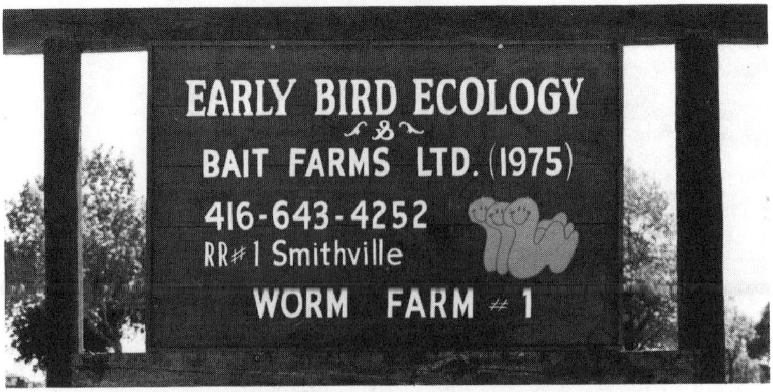

SERVING THE BAIT AND WORM CASTINGS MARKET SINCE 1975

WE OFFER

- ***Giant African Nightcrawlers*** for Sports Fishing
- ***Assorted Sizes of Africans*** for Tropical Fish, Turtles, Snakes, Zoos, and Trout Fishing
- ***Red Worm Breeders***
- ***Organi Green™ Earthworm Castings,*** Top Quality Casts for Organic Fertilizer, Soil Conditioning, and Turf-Top Dressing
- ***Organi Green™ Pro Mix,*** a nutritionally Balanced Mixture of Castings and Peat Loam for all Gardening Needs
- ***Dehydrated Nightcrawlers,*** Our Scientific Method of Dehydrating earthworms insures that they retain an extremely high quality and quantity of Protein. For Supplementary Livestock and Poultry Protein, Fish Hatcheries, and for Amphibians
- ***North American*** and Off Shore Orders Accepted

WE ALSO OFFER

- ***A Successful Growers' Program*** for Raising African Nightcrawlers
- ***A Sight and Sound Video*** of Our Operations; VHS
- ***"Raising the Lively Reds"***—an Illustrated Manual for Cultivating Redworms
- ***Worm Logo*** T-Shirts, Sweats, Golfshirts, Caps (all sizes)
- ***"Adventures in the Underworld",*** a whimsical story for children about Herma, a fun-loving little earthworm. Enchanting illustrations!
- ***The Famous*** KITCHEN KOMPOSTER™, a complete System of indoor Composting
- ***Year 'Round Service • Free Brochure & Price List***

Please call or write us for FREE BROCHURE and PRICE LIST!

Early Bird Ecology & Bait Farms Ltd.

R.R. 1 • SMITHVILLE, ONTARIO • LOR 2AO CANADA
Telephone (416) 643-4252 or (416) 643-4251

George Mallette, President
Anna Scott Draper, General Manager
Charles Mitchell, BSc, Biologist

RAINBOW WORM FARM
24700 County Road No. 95
Davis, CA 95616
Phone: (916) 758-9906
Fax: (916) 756-0414

REDWORMS
WORM BEDDING
POWDERED OYSTER FLOUR
NUTRITIOUS WORM FOOD
BAGGING MACHING & SEALER
CASTINGS
POTTING SOILS
SHIPPING CARTONS
WORM HARVESTERS

The **RAINBOW WORM HARVESTER** is a complete harvesting machine. Heavy duty 16 gauge galvanized steel. No screws to remove in order to change screens. Quick removal of old screens and replacement of new screens in total of 15 minutes. Heavy duty frame is offset to eliminate shoveling material back into your beds. The **RAINBOW WORM HARVESTER** harvests cleaner and faster.

WHOLESALE — RETAIL — EXPERIENCED EXPORT SHIPPER

For free catalog, send self-addressed,
stamped envelope.

ATTENTION FISH BAIT AND GARDENING INDUSTRIES!

Sealright Nestyle® moisture-resistant containers are specially designed to meet the needs of today's earthworm industry.

Easy filling, stackable Nestyle containers have tight-fitting covers that provide maximum ventilation.

Nestyle containers come in two convenient sizes, the profitable pint size with thermoformed plastic cover or the versatile half-pint size with a coated-paper cover.

Ideal for transporting earthworms and all types of fish bait.

SEALRIGHT®

For price information contact: Gene Artz
Sealright Packaging Company
314 South First Street, Fulton, New York 13069
(315) 593-5311 FAX: (908) 938-5389

Worms-A-Lure

© 1993

Specializing in worm bedding. Finely shredded paper with lime; clean, vertually ordor and bug free, PLUS its their food! Just soak in water - ready to go... ***Recycle America***!

25 lbs	$10.00	Plus postage and
Ton	$400.00	handling or pick
1/2 Ton	$250	up your order.

Our ***Red Wigglers*** have allure...

Bed Run	$50.00	5,000 (min. order)
Breeders	$75.00	5,000 (min. order)

If you have any questions, please feel free to contact us at our **NEW** phone number and **NEW** address below:

Connie & John Focht • **194 Ansel Rd.**
Geneva, OH 44041 • **(216) 466-4797**

RED WORMS
PREMIUM BEDRUN

- Affordable Prices
- Live Delivery Guaranteed
- We ship Year 'Round
- We Pay Shipping Costs

STAPLETON BAIT FARM

Rt 1, Box 105
Lumber City, GA 31549
912-363-4451

Hogan's

RD1 Box 1032
Clarendon, PA 16313
814-723-8417

Red Worms

Bait Size
Maximum
800 to the lb. $24.00
Breeders
Maximum
950 to the lb. $18.00
Young Worms
5,000 – $50.00
10,000 – $90.00

Shipped Post Paid
FREE raising instructions with Breeders and Young Worms

OPEN 7 DAYS A WEEK *LIVE BAIT*

SERVING
THE SURROUNDING AREAS

Mail Orders Shipped Anywhere

Live delivery guaranteed!

Redworms • Bait • Breeders • Bedrun

Crickets • Pinheads • 1/8 inch • 1/2 inch • 3/4 inch • Full Grown

Mealworms • Small • Med • Large • Fishing Size • Giant Mealies

WAX WORMS • SPIKES • MOUSEES

Wholesale Tackle • Lures • Spinners • Jigheads • Hooks • Nets

Cupped or Bulk, Packaging Products Supplies

Canadian Nightcrawlers • Minnows • Leeches • Crayfish

Wholesale • Retail • Mail Order • Local

For price information, write or phone:

NIGHTCRAWLERS
CRICKETS
WORMS

SUPERIOR BAIT FARM
7701 Silver Lake Rd
Wolverine, MI 49799
(616) 525-8137

Wax Worms
Mealworms
Giant Mealworms
Spikes (in season)
Mousees (in season)

AUSTRALIAN

Freshwater Marron Lobsters
Yabbies - Red Claw

(Crayfish, Crawfish or Crawdads)

Tremendous opportunities to grow in fresh water lakes, ponds, streams and even back yard pools. Excellent fast growers and ***Big Money Makers.*** *The best tasting lobster in the world!* Sweet as pure sugar. Can grow up to five lbs. each. One female can produce 700 young each year. Send $10.00 for brochures, mailing airmail and handling. Credit for this amount against 1st order.

If you want to get going in this tremendous opportunity ***fast,*** send in $100.00 for a copy of our giant 750 page lobster growers manual. Credit against first order for live Australian Lobsters.

<u>New - Video</u>

V.H.S. only one hour. Demonstrating the techniques currently being used for growing Marron, Yabbies, Red Claw and Giant Australian Crayfish. Only $200.00, includes airmail postage.

HY HUNTER INDUSTRIES

Australian Freshwater Lobsters
Building #707
12001 Highwater Road
Granada Hills, CA 91344
Ph. #(818) 363-0202

Red Wigglers Are the Best

DeArmond's Worm Farm

Since 1987

3923 Greenback Rd.
Utica, KY 42376
502-785-9137 Farm • 502-785-9036 Res.

• Wholesale **• Retail**

• Red Wigglers •

• Bait **• Breeders** **• Bedrun**

• Breeders – 1,000...$14.00

We Offer – FREE information, 100% Guarantee

- Worm Fam Supplies
- We ship 5 days a week
- Kentucky's largest grower!
- Full count and live delivery
- Monthly contracts available
- We will answer your questions (on worms)
- All bait and breeder worms are hand counted
- Supreme Worm Food™ (simply the best)
- Worm shipping bags (cloth, not paper)
  The strongest most durable bag on the market. We make our own!
- Add 10% Shipping to all orders

ATTN: Earn Extra Cash!!
We now buy native earthworms (Dugworms) for our local market. Call or write for current price and information.

New Starter Kits *#1 Selling Kit!*

- 1,000 Red Wigglers Breeders
- 10 lb. Supreme Worm Food™
- 10 lb. Peat Moss
- 128 Page Instruction Book
- Feeding Instruction Card

Only $32.00 or 2 for $58.00
Order your's today!

ATTN: Bait Shops

Prepackaged Worms
Now available
Case – 50-16 oz. cups
60 count worms
Case price – $34.00
Plus 10% shipping
Lowest price around!

Kentucky Crawlers available Spring of '94. 4" to 8" worms
Worm casting special! 50 lb. bag only $5.00 plus shipping

Don and Sherri DeArmond – Owners

Red Worms

ATTENTION –

• Bait Dealers • Start-Up Growers

We Offer –

Red Worms
Native Nightcrawlers, Leaf Worms
Wax Worms, Spikes
China Red Worms

Wholesale and Retail
Breeders, Bait-size, Bedrun
Starter Beds

Experienced in shipping large quantities
Live delivery guaranteed

Worms for –

Fish Bait
Breeding Stock
Growers Plans

Soil Enrichment
Mini-Farms $50 & Up

Sell Worms 24 Hours a Day with Your Own Live Bait Vending Machine

Worm Food (special mix)
10 lb. bags, $1.00 per lb. prepaid
Heating Tapes
Potting Soil
Fish Food for tropical fish (with dried worms)

Finest selection of earthworm supplies & books
BEDS—Growers wanted
BOOKS, Misc. supplies
CUPS—Feed

16 oz. Cups with Vented Lids

$78.00 per 1,000 plus shipping
2,000 minimum
Other size cups and lids available

Bait Cups — Bait Bags

Especially developed for shipping live worms successfully

Worm Sorter / Harvester

Electric powered, portable
Harvests and sorts for size

Earthworm Castings

Grow Northern Nightcrawlers
Send for information

No order too large or too small. We ship anywhere, any time of the year
Prompt, dependable service

For price information, write or phone
Send $1.00 to cover postage and handling

BAIT BARN WORM FARMS
W8697 Island Road
Waterloo, Wisconsin 53594
Phone 414—478-2053 or 414—478-2004

Cyclone Worm Harvesters

Cyclone Worm Harvesters come in assorted sizes. We have 3 screen 18″ drum with 1/10 h.p. motor, 4 screen 20″ drum with 1/4 h.p. motor and 4 screen 24″ drum with 1/4 h.p. motor. We also have a 5 screen 32″ drum and a 4 screen 32″ drum, both with 1/4 h.p. motors.

Cyclone Worm Harvesters are the finest cylinder-type harvesters available today. They are available in different sizes, and are produced locally and are reliable.

One of the more important features common to all is that the owner can change his own screens by removing a few screws instead of returning the machine to the factory. Once a year the whole machine can be easily fitted to remove only casting, then returned to normal use by a simple change of screens.

We designed our machines so that they would be problem-free loading and operating, plus service and durability.

DEALERS WANTED

If you would be interested in more information concerning our harvesters, or would like a price list, please write. Send a large self-addressed stamped envelope plus $1.00 to cover postage.

Cyclone Mfg. Co.

Worm Harvester

W8697 Island Rd
Waterloo, WI 53594
Phone (414) 478-2053

K & W RABBIT FARM

The Home of Quality Redworms
and
New Zealand White Rabbits

REDWORMS

* BREEDERS – BAIT – BEDRUN
* CUPPED OR BULK QUANTITY
* YEAR-ROUND SHIPMENT
* LIVE DELIVERY GUARANTEED
* WORM CASTINGS

Send SASE for information

K & W RABBIT FARM
P.O. Box 50691, Tice, FL 33905-0691
Ph. (813) 543-2620

Griffin Farms

Hybrid redworms   African nightcrawlers
Arkansas tigerstripes

- *large breeder and bait size*
- *generous overcount*
- *guaranteed live delivery*
- *year-round supply*
- *wholesale or retail*

Ed and Pat Griffin
8401 Crossroads
Benton, AR 72015
(501) 776-1750

- *worm castings*
- *worm food*

waterdogs
1-800-729-7312
"Catch The Fish You Dream About"

HOT NEW
Ice Fishing
BAIT

and Leeches

America's Best Fish Baits

For Big Bass, Walleye, Pike, Trout, Salmon, Musky & Catfish

Available Year 'Round

the BAIT MAN
153 E. Woodlawn Box 1233 · Hinckley IL 60520
(1-800) 729-7312

Brown Nose

1 - 5 Lbs., $12   5 - 50 Lbs., $9

Hand Picked Breeders

2,000 - $36   5,000 - $75

Motorized Worm Tumbler

VCR Tape for the Beginner

Breathable shipping bags

Live delivery – Shipping paid

501-355-2154

Jim Knight

Rt. 2, Box 183

Eudora, AR 71640

Lowest Price for

Premium Quality

RED WORMS

Large Breeder & Bait Size

Hybrid Red Worms

5,000 .... $ 65.00

10,000 .... 125.00

50,000 .... 600.00

Premium Bedrun Red Worms

5,000 .... $ 50.00

10,000 .... 95.00

50,000 .... 450.00

100,000 .... 895.00

Large stock of African Nightcrawlers

Live delivery and generous overcount guaranteed.

Prices subject to change

MAC'S

RED WORM FARM

108 King Ave.

Kingstree, SC 29556

(803) 382-8119

Isaac L. McGill & Sons, owners

TATE'S

WORM AND FERN FARM

Rt 1 Box 503

Chatom, Alabama 36518

205-847-2976

Redworms –

- Bedrun-$10.00/1000
- Breeders-$12.50/1000
- Baitsize-$15.00/1000

Brownnose Nightcrawlers –

- Bedrun only $15.00/1000

Grey Wigglers –

- Call for price quote

Will ship anywhere –

- Mail orders welcome

Live Delivery Guaranteed

Robby or Cory Tate

STEWART'S WORM FARM

Hybrid Red Wigglers

Bedrun:

2,000	$16.00
5,000	$32.50
10,000	$50.00
25,000	$118.75
50,000	$225.00
100,000	$400.00

- Generous Overcount
- Postpaid Priority Mail or UPS
- Live Delivery Guaranteed
- Instructions
- Castings: $1.50 /pound
  Postpaid with 10 pound minimum

Marie Stewart
P.O. Box 188
Springhill, LA 71075
Phone: 318-847-4481

Melton Bait Farm

African Nightcrawlers
Redworms
By: the Cup
By: the Pound

Call for up-to-date prices!
Live delivery guaranteed!

"Melton Bait Farm"
P.O. Box 895
Manning, S.C. 29102
803-478-4585
Downtown Davis Station, S.C.
ask for "Butch"

Monterey Red Worms

Bed Run
2,000 – $15.00
5,000 – $35.00
10,000 – $55.00

Postpaid
Generous Overcount

Monterey Worm Farm
9518 Hwy 129
Monterey, LA 71354
(318) 386-2652

Manchester Worm Farm
"Try Us Once - You'll Be Hooked"

Postal Orders & Personal Delivery
Guaranteed Live Delivery

Wholesale & Retail
Competitive Prices

Year-Round Supply
Super Worms . Jumper Worms
Rainbow Golden Giant Meal Worms
Night Crawlers and more...

1131-0 Tolland Turnpike
Box 178
Manchester, CT 06040
203 647-8067

Order Line:
1 800 497-8067

Plans for a . . .

NEW WORM BED

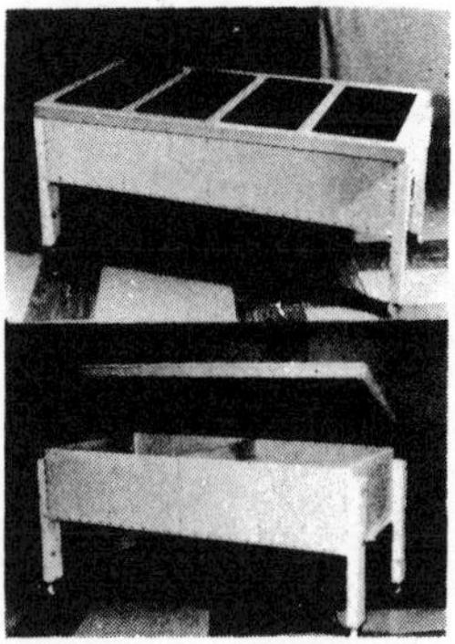

NEW MATERIAL
NEW COVER

RED WORMS – AFRICAN NIGHT CRAWLERS – ALL OTHERS
CHEAPER THAN WASH TUBS
NO MESS – NO LIFTING – NO LIGHTS
CLEAN – ODORLESS – LESS WORK

The plans are easy to READ, lots of PICTURES, a complete PARTS LIST and only the SIMPLEST tools are required.

Whether you're getting into the worm business or are now raising worms, these plans will make your venture more profitable with less work and a lot easier to handle.

CHECK THESE ADVANTAGES

1. NO MESS on the floor.
2. The water that filters through the bed is now used OVER and OVER.
3. NO RUSTING.
4. NO STRAW nor HAY used on top of bedding (with "NO-SCAPE" lid).makes feeding and watering easier.
5. MUCH LESS watering with new lid because of REDUCED evaporation.
6. NO ODOR.
7. NO LIFTING (on rollers). Wash tubs with wet peat moss average 75 lbs.
8. Easy to DOUBLE TIER to save space.
9. One bed equals 3 to 4 wash tubs in area, but takes up LESS ROOM.
10. NO LIGHTS required nor power failures to worry about (with lid).
11. Easy to TURN OVER BEDDING (NO MESS).
12. Dry bedding may be MIXED in worm bed.

The total cost of the bed and plans is LESS than the same area in wash tubs. The price is $6.00 total and INCLUDES the following: (a) plans for NEW worm bed – (b) plans for DOUBLE bed – (c) plans for NO-SCAPE lid – (d) INDOOR raising instructions – (e) three SOILESS bedding formulas.

H and K RESEARCH and DEVELOPMENT CO., INC.

P.O. Box 44184 – OMAHA, NEBRASKA 68144

FRED'S WORM FARM

1409 N. 60th
Lincoln, NE 68505
Fred Patz, Owner

Red Worms

• Breeders • Bed Run

African Nightcrawlers

• Bait Size and Breeders • Bed Run

Starter Kits
Worm Food
Worm Publications

—MAIL ORDER—

Campbell Worm Farm

910-346-4565

Let US help YOU locate or sell Quality Bait and Shipping Supplies

Premium Quality Reds Bedrun
– $10.00 per 1,000
– minimum order of 5,000
Gray Larva Maggots (Spikes)
– $9.50 per 1,000
(when available)
Worm Food
– 15 Lb. $16.00 PPd.
Lime for Beds
– 15 Lb. $13.00 PPd.

Write or call and leave message and phone number.

Prentis & Annie Campbell
207-A Ramsey Dr
Jacksonville, NC 28540

OZARK

RED HYBRID WORMS

• **Bait Size**

• **Bed Run**

POWELL'S

BAIT FARM
Wiseman, Arkansas
72587
Phone:
501/322-7344

AFRICAN NIGHTCRAWLERS

BROWN NOSE WORMS
and
RED WIGGLERS

WHOLESALE

- Cupped or by the pound
- Available year-round

FOOTHILLS BAIT COMPANY
&
RABBITRY

Route 1, Box 936 • North Wilkesboro
NC 28659

Phone: 1-800-487-7878
910-670-9360
Fax: 910-670-9425
Attn: Gene Shepherd

BIG GIANT Worms

Worms big around as your little finger and up to 9 inches long. Tough...Won't break.

Sample $4.00

High velocity blow guns & accessories. Distributors welcome. Add dollars of profit to your bait route.

Swan's Bait Farm

Rt. 1 Box 244
Raymondville, MO 65555
417-457-6215

J. & S. WORM FARMS

"where quality and honesty still exist"

RED WORMS

- live delivery guaranteed
- 3 day shipping within continental U.S.
- year-round shipping
- FREE DELIVERY (250,000 minimum, 250 mile range)
- our special feeding process guarantees you the best worm you can buy!
- shipping bags
- references available on request
- NO ORDERS TOO LARGE!
- generous overcount

PRICES

Bed Run - $5.50 /1,000 + shipping
Breeders - $10.00/1,000 + shipping
J. Downs • Scott Downs
411 S. 4th Ave. • Paragould, AR 72450
Phone (501) 236-7619

Denton's Quality Worm Farm

6441 Byrnes Rd.
Vacaville, CA 95687
707-448-0453

Red Wigglers	**Sale Price**
Breeders & Bait Size	
#5,000	$45.00
#10,000	$80.00
Bedrun	
#5,000	$35.00
#10,000	$60.00

over 10,000 $1.00 LESS per 1,000

Call for Prices on Gray Crawlers and African Crawlers

Worm Castings $1.00 per pound

Live Delivery Guaranteed!

Redworms

Large Bait-Size Hybrid Breeders

1,000 –	$15.00
5,000 –	$60.00
10,000 –	$100.00

Bed Run

5,000 –	$25.00
10,000 –	$40.00

add 10% for postage • Visa/MC accepted

Year-Round Shipment • Live Delivery Guaranteed

Instructions–Books–All Types of Supplies

Finest Quality & Competitive Prices

Smith's Worms

Rt. 2, Box 205A • Boston, GA 31626

Phone 912-498-1605 • 1-800-252-5133

C & G ENTERPRISES

P.O. BOX 658
PERRY, OHIO 44081

Owner: Carol Miller

Phone: (216) 354-5414

"Honest People – Honest Count"

Premium Redworms

Bait Size .... $15.00 per pound

Breeders .... $12.00 per Thousand

Bedrun .... $10.00 per Thousand

"Guaranteed Live Delivery!"

Phelps Worm Farm

Rt. 1, Box 433
Silva, MO 63964
(314) 783-3944

"Raising Premium Redworms"

Wholesale Accounts Welcome
Nationwide Shipping

- *Bait* • *Breeders* • *Bedrun*
- *Bulk or Cupped*
- *Wholesale & Retail*
- *Worm Castings*
- *Worm Books*
- *Free Information*
- ***100% Guarantee***

L & T ENTERPRISE WORM RANCH

119 N. 8th
Humboldt, KS 66748
Ph. 316-473-2022

Wholesale Retail

African Night Crawlers
Breeders or Bait
Breeders Hand-picked
and
Counted

FREE Information

PREMIUM QUALITY

RED WORMS

Bed Run
1,000 - $5.50 + Shipping

Breeder's
1,000 - $10.00 + Shipping

Live Delivery
Guaranteed

$6.00 per lb. + Shipping

Family Worm Farms

1553 Greene 611 Rd
Paragould, AR 72450
501 586-0451
501 586-0568

Southern Worm Enterprise

REDWORMS – BROWN NOSE
GREY NIGHTCRAWLERS – JUMPERS
BAIT – BREEDERS – BEDRUN
CUPPED OR BULK

12118 MARILYN LN.
HAMMOND, LA 70403

504-294-2603

THE WORM FACTORY™

FEATURING:

MINI WORM FACTORY™ - *THE* COMPLETE INDOOR VERMICOMPOSTER! INCLUDES, BIN & LID, BEDDING, REDWORMS, INSTRUCTIONS, AND REDWORMS. ALL YOU ADD IS THE GARBAGE! RETAIL & WHOLESALE PRICING AVAILABLE IN CANADA AND THE USA.

ALSO OFFERED:

- 25+ EARTHWORM BOOKTITLES FOR TEACHERS, FARMERS, VERMICOMPOSTERS, & ALL WORM ENTHUSIASTS.
- BULK REDWORMS - CANADA ONLY

FREE BROCHURE\CATALOGUE FROM - THE WORM FACTORY™, 129S SHERBROOKE STR. E., PERTH, ONT., CANADA, K7H 1B1, 613-267-5540.

HAPPY HOOKER BAIT

B & J Enterprises
HCO 1 Box 17
Pinehurst, Idaho 83850
(208) 682-4223

Live bait & worm bedding
We sell the best

Mature, Bait, Pitrun
Year Around Northern Nightcrawlers
Red Wigglers – Maggots

Mail Order and Local
Wholesale and Retail

Owner:
B. J. Dotson

WORMS 4 U

NIXON & SONS LIVE BAITS

Rt. 2 Box 63
Evington, VA 24550
(804) 821-3191
(Owner Mike Nixon)

CANADIAN NIGHTCRAWLERS
- Bulk Packed - 1000/Box $48.50
- Cupped - 12/Cup, 90/Case .... $86.00 + Shp
- Cupped - 18/Cup, 45/Case .... $55.00 + Shp

RED WIGGLERS
- Bulk Packed - 5000/Box .... $60.00 + Shp
- Cupped - 45-50/Cup, 100/Case $85.00

ALABAMA JUMPERS
- Cupped - 35-40/Cup, 48/Case .... $48.00 + Shp

MEALWORMS
- Bulk Packed - 4000/Box .... $48.00 + Shp
- Cupped - 24/Cup, 100/Case $65.00

WAXWORMS
- Bulk Packed - 5000/Box .... $65.00 + Shp
- Cupped - 24/Cup, 100/Case $65.00

GUARANTEED YEAR ROUND LIVE DELIVERY AND SHIPPING

RABBIT HILL FARM

Brown Nose Worms

Quality Pit-Run
Brown Nose Worms
Post Paid Orders
Wholesale, Retail
Call For Prices

• ORGANIC •

Worm Casting	Produce Plants
Potting Soil	Herbs
Rabbit Manure	Earthworms

Joanne Mertz
J.A. "Jay" Mertz
Rt. 3 Box 2936
Corsicana, TX 75110
(903) 872-4289

WILLINGHAM

Worm Farm

We offer:
Bedrun Red Worms
by the Pound or Ton
for Fish Bait or composting
- $8.00 - 1 lb. plus postage

Year Round Shipment!

Live Delivery Guaranteed!

Alan & Sybil Willingham
Rt 1 Box 241
Butler, GA 31006
(912) 862-5545 1-800-223-WORM

REDWORMS

Large Worms For Bait
Cupped or Bulk

Bed Run

Wholesale

- **Worm Bedding**
- **Mulched Paper**

GOLDWEN ECOSYSTEMS

RR# 1
Paris, Ontario
Canada N3L 3E1
Tel. (519) 756-6469

Live Delivery

LEE'S WORM RANCH

Home Grown

Lee R. Allen
517-893-4119

2619 Bullock Rd.
Bay city, MI 48708

RED WORMS

Call or write for competitive prices on worms & castings.

SPECIALIZING

In municipal yard waste composting.
Tours available on request.

AFRICAN WORMS

Quality Bait

- Wholesale
- Bulk

PLASTIC BAIT CUPS

- 12 & 16 oz. – Plain
- 16 oz. Printed with fishing scene, done in multi-colors with the logo:

You never strike out with worms!

CROSSROAD BAITS

5375 Hwy. 321 North
LENOIR CITY, TN 37771
(615) 988-3853

– Since 1973–

SOLANO WORM FARM

(NOTE: Solano Worm Farm has incorporated.) New name: Vermi Technology Unlimited, Inc. Same *"ole worm farmer"* at the helm guiding it into the 21st. century. V.T.U., INC. will continue offering only quality worms with no increase in price and the best service in the U.S. We continue our commitment to excellence. Thanks.

L. D. Martin
Pres. V.T.U.

REDWORMS: Capsules, Bait Size, Bedrun, Large Bait Size

ALABAMA JUMPERS:

Worm Feed
Castings
Potting Soils
Shipping Cartons
Books
Mini Worm Farm

NEW PRODUCTS & SERVICES:

Worm Harvester – All Aluminum Construction
Worm Grader
Small Vermi Composting Bins for: Home, Office, Schools, etc.
Commercial Vermi Composting Bins: Capacity – 60 to 1000 lbs/day (to -30°)
Consultant Services: Full service from worm farming to vermi composting at landfills.

For **free** catalog, send self-addressed, stamped envelope to:

V.T.U., INC.
P.O. Box 130
ORANGE LAKE, FL 32681

WHOLESALE – RETAIL – Ship the year-round

SPECIALIZE IN EXPORTS
PHONE: (904) 591-1111 FAX: (904) 595-7333

By

Worms Weber

- 25 Years Experience -

- African Nite Crawlers
- Tiger Wigglers
- Baitsize or Bedrun
- Cups or Bulk – Hand counted
- Castings & Compost
- Hand Picked Breeders
- Organic Garden & Lawn Supplies
- **Worm School Taught**

Call (501) 449-5348 or 481-5784
P.O. Box 1126
Yellville, AR 72687

THE BAG LADY

–Best in the Country–

Cool, Breathable, Earthworm Bags. Cut shipping losses in half, ship year-round.

7x12 in. . $0.30
9x16 in. . . . 0.45
11x20 in. . . 0.56
(Minimum order of 100 bags, 250 bags on all custom orders)

Send SASE for Catalog to:

J.A. Martin
18570 NE 43rd. Ct.
Citra, FL 32113

Will custom sew any size bag

Premium Quality Red Worms

Bed Run

$9.00 Per 1,000
Minimum order of 5,000
Live delivery guaranteed postage prepaid

We will answer your questions *(on worms)*.
All Inquiries Welcome.

Lucky Hooker Worm Farm
295 N. State Street
Box 408
Ferron, Utah 84523
(801) 384-2904

LOWEST PRICES FOR PREMIUM QUALITY REDWORMS

Large Breeder & Bait Size

5,000 .... $35.00
10,000 .... $65.00
20,000 .... $120.00

Bed Run

5,000 .... $22.50
10,000 .... $35.00
20,000 .... $60.00

Postpaid • Generous Overcount • Wholesale • Retail
Live Delivery Guaranteed • Year 'Round Supply
FREE Information • Castings & Prepared Soils

Wormco

PO Box 338 • Lincoln, CA 95648
(916) 645-9419

CAIN'S WORM FARM

Redworms **Castings**

Live Delivery Guaranteed

5678 Weber Rd
Dixon, CA 95620

P.O. Box 1155
Davis, CA 95617

Phone: 707 448-6836

WOEHRLE'S WHOLESALE BAIT & SUPPLIES, INC.

1-800-227-BAIT

95 CHURCH AVE MT POCONO PA 18344

WE SPECIALIZE IN LIVE BAIT & SUPPLIES
FOR THE BAIT DEALER
WHOLESALE ONLY
WE GUARANTEE LIVE DELIVERY
AND WE SHIP ANYWHERE
CALL FOR FREE CATALOG

TOP QUALITY CANADIAN NIGHT CRAWLERS
CALIFORNIA RAISED MEALWORMS
DUG WORMS (TROUT WORMS)
WAXWORMS
SPIKES
MOUSIES
SUPERWORMS

TANK FILTERS
AERATORS
MINNOW BUCKETS
BAIT CHEMICALS
CHILLERS

ROM. 10:13 'WHOEVER CALLS UPON THE NAME OF THE LORD WILL BE SAVED'

EARTHWORMS & SUPPLIES

RR3 BOX 291 – OWENSVILLE, MISSOURI 65066-9549
Phone 314-437-3605

EARTHWORM PACKING & SHIPPING SUPPLIES

Bags – Miracloth
Bag closer tool & clips
Bag heat sealer
Bag ties
Shipping boxes – all sizes
Packing tape dispenser & tape
Scales – 5 lbs., 50 lbs.
Scales, electronic – 5 lbs., 10 lbs.

MISCELLANEOUS EARTHWORM FARM SUPPLIES

Bait buckets, 1/2 gal. personal use
Earthworm buckets, 5 gal., escape-proof
Earthworm bedding
Earthworm books
Earthworm feed grinder
Earthworm bins – metal or plastic
Earthworm separators – (separates worms from bedding)
Earthworm feed – special mix
Earthworm cups, "foam" 12 oz.

EARTHWORM OFFICE SUPPLIES

Printed items – stationery, shipping labels, business cards, price lists, etc.
Postage scale – ounces

"There's Money In Earthworms" (book), $1.00

WORM WORLD

Pennsylvania's Finest Redworms

Quality Breeders and Bait Size

1,000 . . . . $15.00
5,000 . . . . $55.00
10,000 . . . . $100.00
50,000 . . . . $400.00

Guaranteed Live Delivery
Generous Overcount
Postpaid

Kay & John Ihnat
(412) 356-2397

R.D. #2 Box 188
Ihnat Lane
Avella, PA 15312

WORMS

EARTHWORMS FOR:

- Gardening
- Composting
- Raising
- Fishing

Redworms – Large:

1,000 ....$12.95
2,000 ....$21.00
5,000 ....$47.50

Bedrun:

5,000 ....$22.50
10,000 ....$37.95

Gray Nightcrawlers:

500.........................$23.95
1,000 ....$39.00

Postpaid • Instructions

BRONWOOD WORM GARDENS 5P

BRONWOOD, GA 31726-0028
PH: 912-995-5994

REDWORMS

Bedrun

5,000 ....$21.75
10,000 ....$34.50
20,000 ....$66.00

- Live Delivery Guaranteed
- We pay shipping costs
- Instruction shipped with worms

Boot's Worms Farm

Rt. 7 Box 575-A
Lumberton, NC 28358
Ph. 910-738-7432

Regional Bait & Marketing

Premium, Quality Redworms
Year 'Round Supply

* Large Breeder & Bait Size

1,000 .... $10.50
5,000 .... $42.50

* Bed Run

5,000 .... $22.50
10,000 .... $40.00

* Worm Castings .... $1.00 LB.

DON & GARY WOLCOTT
8006 Plank Rd. Thompson, OH 44086
CALL: 1-800-RED WORM (733-9676)

Tell Advertisers You Saw It In The Buyer's Guide

CANADIAN RED WIGGLERS

NOW available
to our Canadian customers
BEDRUN
and compost starters

Will Post or UPS
Prepaid containers...$15.00
approx. 1000/package

1-519-664-2701

Paul Fretz Enterprises Inc.

Operating Bunnyview's
Commercial Rabbit Ind.
Route 2, West Montrose,
Ontario NOB 2VO
Canada
Fax: 519-664-1492